2023 MOONLY

Course to 33

Marcos Cervantes Janssen

2023 MOONLY

Course to 33

By: Marcos Cervantes Janssen

First edition: November 3, 2022

Copyright © 2022 Marcos Cervantes Janssen

Edited by Editorial letr@Roja

https://newtekjanssen.es.tl/letra3roja@gmail.com

https://www.facebook.com/LETRA3ROJA

https://www.newtek.janssen @gmail.com

https://payhip.com/letra33roja

INDEX:

- PROLOGUE: 5
- ORBIT: 7
- GRAVITY: 9
- MOON: 11
- GALACTIC EXPANSION: 13
- 2023: 15
- 2033: 16
- RACES: 18
- QUANTUM TECHNOLOGY; 20
- EPILOGUE; 22

PROLOGUE:

Living on our moon is the perfect training, to start the experience in the expansion of our habitat outside the planet, for that to create the necessary conditions for our life, already established outside the earth, it prepares us not only physically, our scientific knowledge of how to live under these biological circumstances is a drastic change, but even more so our psyche must have even more adaptability with all the necessary requirements. As a human race, we are waking up to cosmic consciousness, and we have become aware of the animal reality, which is definitely not ours as humans, we have adapted to this planet but we are in the full moment of understanding that due to the requirements natural to our being, we always go beyond instinct.

As we already know, the Moon is the prelude to Mars, the trip designed by Germany to Mars was calculated with the inclusion of the lunar arrival, in these 50 years, the Moon is our base for intergalactic launches, it is therefore the border of our first sky, I describe it as the last terrestrial continent for intergalactic transfer, today the moon, will no longer be a mystery if not quite the opposite, its dark side being our observatory par excellence. We live not the collapse of humanity, but the beginning of our true nature as the race that we always were, visitors to this beautiful planet that we must heal, because our true capacities allow it, we have sufficient evidence at this precise moment of the technological capacity that has been developed to provide a real solution to these problems that we ourselves have allowed.

ORBIT: To orbit

an idea, is to have it close to our immediate thoughts in order to develop it, with practical, immediate and objective actions. The space race is not only a historical event, orbiting ideas and goals for decades brought with it increased confidence in ourselves as humanity and thus the proposal to orbit the moon was a project that through the first launches in germany and russia was thus possible, leaving the planet to find the orbit to be conquered by the first human rockets, the orbit is that precise distance of balance between gravity and forces of attraction between the earth and the sun. An equidistant and stable balance to maintain without the need of any force a circular route, around the planet, in full rest and balance.

It is very important for us as human beings to understand that each advance in the space race is to rediscover the paths of our origins, the orbit defines time, constant cycles that govern planetary history for centuries. The orbit in general is circular or elliptical, the celestial bodies orbit each other after the beginning of each universe through a circular birth, this is how the orbit describes the constant path when there is no interference in its path. In the stars, the orbits always trace a spiral figure, thus traveling systems through space, having orbits for galaxies and nebulae that all travel in infinity. In the field of human psychology, our life traces an orbit that if we know it, even in the cyclical monotony of events, it is how we advance in life, traveling in space also teaches us to evolve.

GRAVITY:

Gravity is a constant force, of universal order; gravity is a geomagnetic force, which forms a dynamic network throughout the universe. If we are able to calculate the forces between all the heavenly bodies, we will understand the destiny and origin of everybody in the vast and infinite space.

Gravity is a structural form throughout the universe, which changes over time, the existential energy that makes up the mass is permanently affected by it, giving rise to orbits and constellations that we have observed and studied for centuries. Weight is a phenomenon derived from gravity, the formation of planets and their movements go hand in hand with this impressive force, which is provided with infinite casualties and defines the galactic destiny eternally.

In the social aspect, we find a gravitational force within our behavior; the attraction of philosophies or doctrines have governed the destiny of history for centuries; the rotation of events that we repeat century after century, but at different times. The seriousness of the matter, is an expression to point out and highlight a particular event, for a particular result, gravity is a transforming force in the field of the solution, example is the birth of a plant, or the flight of a drone When fertilizing the queen bee, gravity requires an effort on the part of the person who crosses this force, that is how we come to understand that the problem also carries the solution within.

Gravity, as part of the existential equation, reveals to us the origin and destiny of the movement and perpetual dynamics of the universe, that transformation without any loss, called perpetual destiny.

MOON:

The moon is our companion satellite, its gravity is of vital importance for our planet, its effect on the waters, and the night light it provides gives balance and stability to the different vegetable and mineral ecosystems of our planet, on the other hand in the psychology of the inhabitants of the planet the moon exerts a face in their philosophy, beliefs and spiritual human world, based on the history affected by the presence of this beautiful star, our moon. Whatever the material constitution of the moon, it is impressive, also its spherical shape, it is perfect and its rotation is unique in the galaxy. Its crustal solidity reveals a resistant anatomy, its perfect movement leaves an advanced civilization in sight, transforming our planet into the only habitable one in our galaxy.

Whatever the origin, purpose and mission of the moon, by the fact of existing, it is said that we exist as a human race protected and provided with great advantages by this satellite, in perfect orbital movement, exact size and distance for unique eclipses in all the universe. Each measurement in diameter, or distance, as well as speed and time, entail a constructive harmony, and conducive to the expansion of life in the universe, the moon is therefore an object of study, and a wonderful star which allows us to know the character and movers of our system, we must understand that just admitting the possibility of a moon placed for us, entails the responsibility of advancing, to the point of being able to put in orbit, moons on the planets to be terraformed, thereby creating the geomagnetic forces that They provide optimal properties to the water for crops.

GALACTIC EXPANSION:

This is how the moon's purpose is to be the object of revelation to our race in order to advance in the intergalactic expansion of our species. We know that our human life does not allow us to contemplate true evolution in all its forms, but thanks to historical records, we have the opportunity to understand through study how great our habitat and life is in this galaxy. We are a growing race by reproductive nature, and it is natural that if space expands eternally, so do we, we have the genetic instruction to reproduce ourselves and even more so to preserve our life permanently, thus we will form as individuals a community with no growth limit.

We have not confirmed that we are an endangered breed, but every day we experience a unique experience of temporality and effort to increase our presence, in the best possible way. Several satellites travel through space, fulfilling their different missions, he thus mentioned the moon as the planet's own satellite in transformation, every human being desires within himself freedom and an opportunity to manifest everything possible. It is important to understand that expansion always requires communion between human beings, this commission promotes agreements and cooperation between the different movements of all kinds, to be inclusive with integrity, the union always fosters multiplication and expansion potential through natural guidelines , endless expansion is already in our genes.

2023:

This Year is the official beginning of the lunar knowledge exposed to humanity, its true constitution, who has truly studied it and that has been achieved in this last half century on its surface, in the true official beginning of 2033 Martian, here in the lunar orbit where we have all the potential and training to forge ourselves as a race that is integrated into intergalactic society, this entails a change of consciousness and a devastating opening of conditioning by radical conservatives. It is in this year that we are born as a race of humans in true integral awakening. No movie and no story can reveal how difficult but wonderful it would be to transform the entire planet as we expand in every possible way thanks to this new reality.

2033:

In reality, Mars was the inhabited planet before Earth, and Venus the possible inhabited planet in the future. the galaxy expands, in such a way that, the third place that the earth occupies with its moon, is the optimal place for life. The distance between the sun and the planet, plus all the special specifications of the earth, makes it the only habitable place for the moment in our known galaxy. Venus would need to have a moon, which it does not have, one would have to be placed to generate the necessary forces on the planet for its correct terraforming. Placing a satellite in orbit requires a lot of energy and cosmic architecture engineering, this is how through the study of our ancestral past, we can discover our origins, and learn not to repeat mistakes.

We are on our way to 2033, this will be the Martian year in which our ancestral past will be better known.

Mars has an archeology to discover, and what we have managed to position on its surface, it is very important that it continues to improve, thanks to the moon and the knowledge that we have due to its proximity, the means and knowledge have been developed to coexist outside of the planet with what is abroad, we must recognize and accept that our knowledge can only be established in the correct way, the more we are abroad, for this it is necessary of all the passes and all the experiences already lived by those who know the subject, Mars is very useful to value and preserve our present home, the earth, which together with the moon, form the exemplary gravitational system to follow, for the geoformation and restoration of any other exterior.

RACES:

RED, BLACK, YELLOW AND WHITE, are the 4 races that are spoken on earth, each with very special characteristics as well as different from each other, the human race is one, and within this group the sub Divisions are marked by physical and mental attributes that generally identify each one. Each region with its variety of climates greatly influences the formation of your personality, within the space race you have all the races working in fluid and respectful communion. The experience of leaving the planet has taught each country to respect each other for belonging, humanity as a symbol of planetary union. Preparing you with this to the open and public coexistence with more races of our galaxy or neighbors, this will be the absolute planetary difference.

We are on the verge of a complete change of reality in our planetary history. Our consciousness has awakened to new possibilities, however, we should never be guided by fables or myths, we will walk from 2023 to 33 in a reality already expected but until now experienced, current science has reached a point of quantum experience, that is why magic and superstition will be disproved by knowing the reality of our entire galaxy and not just the events of our little planet. coexisting with all the races of the planet, will train us to be inclusive, patient and cooperative. So this is for the common good, it requires caring for and valuing our planet, and all life existing on it, in the same way in this our universe, they will only deal with adequate species for the understanding of cooperation and mutual learning; this is how accepting every race is to evolve.

TECHNOLOGY QUANTUM

In this last decade, the loop theory brings us to a new era of quantum technologies, let us remember the subject of fractals, and fractional numbers, all this reveals to us the vast field of development in the subjects of space and energy, directly involving time, no longer as a constant, but as one more variable for the expansion of results at the quantum level, this is actually the mathematical way of describing different dimensions that cohabit and share characteristics and functions in common. This theme takes the real possibility of bending space through time, so we can work with energy-mass at the will of each galactic experience, quantum physics opens the doors of lost solutions and today we will launch a generation that awakens.

The possibility of traveling through temporary doubles in space, which this new area offers, is just discovering ancient ways of interacting with the rest of the universe. Our planet will find beauty and value, when we are in another place remembering our first home, the planet and its great diversity. As mortal and temporal beings, we will value and gladly take back what we already have. At the quantum level, being in several places at the same time was achieved by the fragmentation of time, that is why the only way to know and understand black and white holes will lead us to the technology that will move the planet to a promising future. and without the need for violence because it is a real liberation, quantum physics is the mental interaction of already existing matter, with it we will understand who we are and what we must do to evolve.

EPILOGUE:

I conclude this essay, reaffirming the present existence of a reality to be revealed to the whole society, today all the terrestrial and visiting races will understand the union of our origin, so different in the details, more intimately similar in our interior. The moon and mars are the first human project for intergalactic citizenship, Venus a future, and the quantum is the instrument revealed to humanity by the infinite and eternal universe in which we inhabit, live and move. Every day that passes is one more step for the great universal reconciliation in favor of order, beauty and total discipline, only together and truly free, our minds can be in the same creative frequency throughout our planet, materially we are equal and brothers .

WE WERE BORN BEING IN REALITY WHO WE ARE, LET'S REMEMBER AND ACT NOW.

All rights reserved. Under the sanctions established

in the legal system,

without written authorization from the holders of *Copyright* ©

the total or partial reproduction of this work by

any means or procedure

, reprography and computer processing

.

Hello, I am a researcher, writer and communications engineer, throughout my life, I have experienced strong situations in every way, I wish that your life goes better and better, and that you evolve as much as you can by expanding your knowledge, mind and will, I am sure we can find an expand our existence, I want to accompany you always, and I thank you in advance YOU ARE

Living on our moon is the perfect training, to start the experience in the expansion of our habitat outside the planet, for that to create the necessary conditions for our life, already established outside the earth, it prepares us not only physically, our scientific knowledge of how to live under these biological circumstances is a drastic change, but even more so our psyche must have even more adaptability with all the necessary requirements. As a human race, we are waking up to cosmic consciousness, and we have become aware of the animal reality, which is definitely not ours as humans, we have adapted to this planet but we are in the full moment of understanding that due to the requirements natural to our being, we always go beyond instinct.

www.ingramcontent.com/pod-product-compliance
Lightning Source LLC
Chambersburg PA
CBHW050328220526
45465CB00005B/2190